OXFORD
UNIVERSITY PRESS

Oxford New York

Auckland Bangkok Buenos Aires Cape Town Chennai
Dar es Salaam Delhi Hong Kong Istanbul Karachi Kolkata
Kuala Lumpur Madrid Melbourne Mexico City Mumbai Nairobi
São Paulo Shanghai Taipei Tokyo Toronto

Published by Oxford University Press, Inc.
198 Madison Avenue, New York, NY 10016
www.oup.com

Oxford is a registered trademark of Oxford University Press

Library of Congress Cataloging-in-Publication Data is available.

ISBN 0–19–521993-7

1 3 5 7 9 10 8 6 4 2

Printed in Italy

Acknowledgments

The publishers would like to thank:
For the Science Museum: Peter Davison, Sam Evans, Douglas Millard, David Mosley, John Robinson

All photos reproduced in kind permission of the Science and Society Picture Library
with the exception of the following:
Allsport; p19
Britstock; p14c
Castral; p20-21t
Digital Vision; p9tr, 11b, 13b, 14c, 17tr and br, 18bl, 20bl
Eye of the Wind; p7
Frederic Clement; p21cl - Every effort has been made to contact the copyright holders
but this has not been possible. If notified the publisher will be pleased to rectify any error or omissions.
Japanese Embassy; p9br

speed

Contents

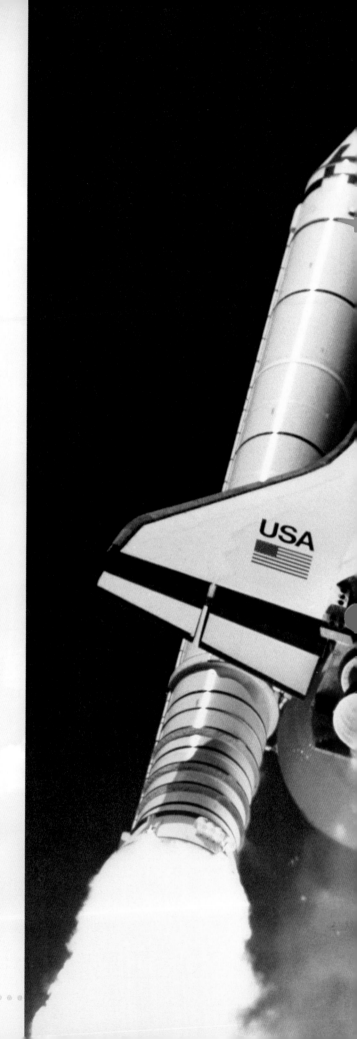

Sailing at speed

SINCE THE EARLIEST times, people have wanted to travel quickly. Horses were used as a swift mode of transport, but they tired easily. For longer distances, the fastest way to travel was often by boat. To get up to speed, you needed either lots of people rowing, or big sails that could be adjusted to catch the wind.

⬆ Swinging sail
About 4,000 years ago the Egyptians built sailing boats to travel along the River Nile. The crew could adjust the angle of the square sail to keep up the speed.

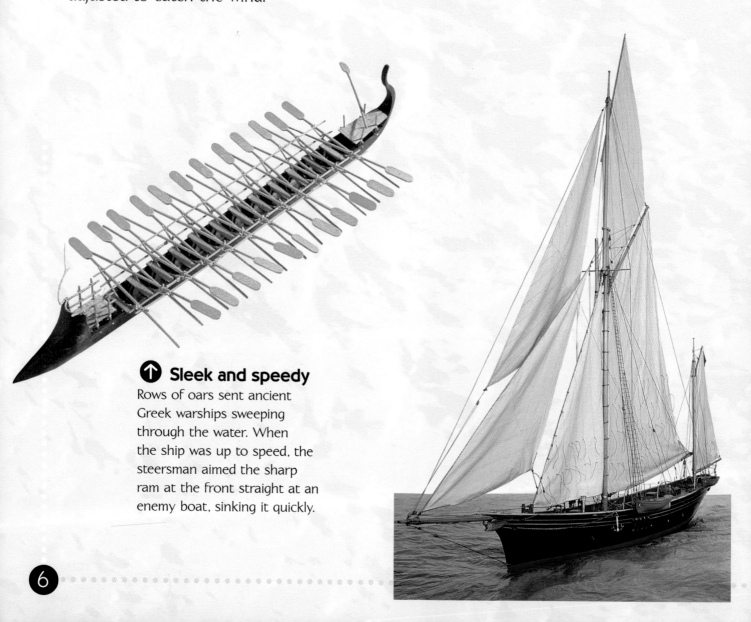

⬆ Sleek and speedy
Rows of oars sent ancient Greek warships sweeping through the water. When the ship was up to speed, the steersman aimed the sharp ram at the front straight at an enemy boat, sinking it quickly.

➡ Sail power

In the 19th century, shipbuilders developed really fast sailing craft known today as tall ships. These could sail from Australia to Britain in 80 days—a record at that time. They got their speed by carrying lots of huge sails on their tall masts—some ships had almost 30,000 square feet (2,970 square meters) of sail. The fastest tall ships of all were the clippers, which carried valuable cargoes, such as tea, from China to western Europe.

⬅ Racing yacht

Yacht racing became popular in the 19th century, when vessels such as this one, the *Jullanar*, were built. They had slender, lightweight hulls, to slice their way through the waves.

Steaming ahead

FOR OVER 150 YEARS, trains have been our fastest means of transport on land. With more powerful engines and straighter track, they are getting swifter still. The most exciting development is the experimental maglev train, which "flies" above its track at incredible speeds.

↗ Rocket

The steam locomotive *Rocket*, designed by British engineers George and Robert Stephenson, was fast for its time. In 1829, it reached a speed of 29 mph (47 km/h)!

↗ Streamlined speedster

In the 1930s, two British railway companies, the London, Midland and Scottish (LMS) and London and North Eastern Railway (LNER), competed to run the fastest passenger trains between London and Scotland. In 1938, the sleek, streamlined *Mallard* of the LNER achieved a speed of 126 mph (203 km/h). No other steam locomotive has ever broken this record and many modern diesel and electric trains don't have such a high top speed.

⬆ Smooth speed

The French TGV (Train à Grande Vitesse, very fast train) is one of the world's fastest trains. The electrically powered train can travel at up to 320 mph (515 km/h). At its service speed of 180 mph (290 km/h), passengers get a smooth, stylish ride.

⬆ Flying along

Maglev trains use magnetic force to hover slightly above the track. With no wheels to rub against the track, maglevs can really zoom—at up to 343 mph (552 km/h).

Built to race

AS SOON as the first cars were built, at the end of the 19th century, many drivers wanted to be the fastest. At first they raced ordinary cars, but soon people started to design cars specifically for racing. They made them faster by streamlining the bodywork and fitting special engines. Soon there were all sorts of races for different types of car, from small sports cars to hugely powerful Formula One racers.

⬆ Starting line

Drivers wait for the start of a race at Britain's Brooklands circuit in 1927. The race was won by H. W. Purdy driving number 10, one of several cars made by the Italian manufacturer Bugatti. His average speed was 55.47 mph (89.26 km/h).

⬀ Silver Arrow

By 1938, when this Mercedes was built for the Silver Arrows team, racing cars were streamlined and the driver sat low down in the cockpit. The car's slender shape helped it to reach a top speed of around 200 mph (322 km/h).

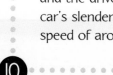

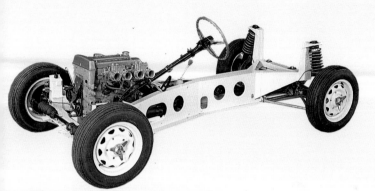

↗ Car with a backbone

The 1962 Lotus Elan had an unusual construction, based on a "backbone" made of steel. This skeleton gave the Lotus strength, while a fiberglass body (not shown) kept it light, so that the small engine could power the car to 120 mph (193 km/h).

↓ Formula One

Today Formula One racing is enormously popular—and incredibly fast! The design of the open cars is carefully controlled to keep them safe and to make sure no team has an unfair advantage. With their broad tires, Formula One cars grip the road firmly. Their bodies are designed to create a vacuum effect under the vehicle, forcing moving air under the car to "suck" it down toward the road. These mean machines can reach top speeds of up to 190 mph (306 km/h)—but the curved race tracks slow them down.

Jet propelled

THE JET ENGINE was invented by British and German scientists, working separately, during World War II, when both sides wanted to produce faster, more efficient fighter planes. Jet aircraft were not much used in the war, but today jets power many airplanes, from fighters to airliners.

➔ Trendsetter

With an engine designed by inventor Frank Whittle, the Gloster E28/39 was the first British aircraft to be jet propelled. It made its first flight in May 1941. Whittle had spent about 12 years working on jet engines, often on his own and with little money to pay for his research. The Glosters flew well, and by 1944 Britain's Royal Air Force was using the Gloster Meteor. Meanwhile, the Germans had developed their own jet, the Messerschmidt Me 262.

⬅ Jet engine

Many jet engines have a large fan at the front to compress air and suck it in. Fuel is mixed with this air and burned, producing a stream of hot, expanding gases. These gases stream out of the back of the engine, pushing the airplane forward.

⬆ Jump jet

The Hawker Harrier "jump jet" is a vertical take-off and landing (VTOL) aircraft. The nozzles of its jet engine can be turned toward the ground. This gives the airplane vertical thrust, so it climbs straight up into the air without needing a runway.

➡ People's plane

Today all large airliners have jet engines. Some of them can fly at over 620 mph (1,000 km/h). Their speed and economy have brought fast air travel to millions of people, making the world seem a much smaller place.

Through the sound barrier

BOOM! An aircraft above us breaks the "sound barrier." Sound moves through the air at 736 mph (1,188 km/h). For years people dreamed of being able to travel this fast, but the huge stresses and strains at this speed tore ordinary aircraft apart. Then, in 1947, an American pilot named Charles Yaeger succeeded for the first time. Today, with better aircraft design, people regularly travel faster than the speed of sound.

→ High flier

The X-15 aircraft was built in the United States in the 1950s for research into high-speed flight. Its pilots discovered what it was like to fly miles above the Earth's surface, zooming much faster than the speed of sound.

↖ Fastest for passengers

Concorde was the world's only supersonic (faster-than-sound) airliner. Its top speed is 1,450 mph (2,333 km/h), about twice the speed of sound. Its unusual design, with slender body, triangular wings, and pointed nose, had to be tested for years using models (below) in a wind tunnel. Despite one fatal accident in July 2000, Concorde was a great success, regularly crossing the Atlantic Ocean in 3 hours 30 minutes.

← Blasting into space

To blast into space, the Space Shuttle needs huge
power. This is because it has to reach an incredible
17,440 mph (27,904 km/h) to counter the
downward pull of the Earth's gravity. The power
for this speed comes from the Shuttle's main
engines and two strap-on rockets. The
rockets are discarded and fall back to
Earth when empty and can be reused.
The Shuttle engines are fueled by
a massive tank that is also
jettisoned when empty and
allowed to burn up in
the atmosphere.

Power on water

ACHIEVING SPEED

on water is all about shape, weight, and power. Modern designers use the latest materials, such as plastics and carbon fiber, to make their craft streamlined, light, and strong. Then they add powerful engines to make them whizz through the water.

⬈ British racer

The elegant speedboat *Miss England* was built in 1929. Designs borrowed from airplanes made the lightweight hull, allowing the single engine to power the boat at impressive speeds. In *Miss England*, British powerboat racer Henry Segrave won the World Motor Boat Championship at Miami in 1929, reaching a record-breaking speed of 91.9 mph (147.9 km/h).

⬅ Bluebird

The jet-powered boat *Bluebird* belonged to British record-breaker Donald Campbell. Campbell was determined to be the first to travel at over 200 mph (322 km/h) on water. He achieved this in 1955, broke his own record several times, but was killed in a crash in 1967 while trying to reach 300 mph (483 km/h).

→ Streamlined for success

Both water and air produce drag, the force that slows down a moving object. Racing boats are streamlined, with curved, flowing bodies just like sports cars, to reduce drag as much as possible.

AND

↓ Flying along

This modern offshore racing powerboat has a very slender hull. It is designed so that when the boat goes fast, the front lifts right out of the water, reducing drag and making it super-speedy.

Fast bikes

IN A BICYCLE RACE, both rider and machine have to be ready for speed. That means hard training for the competitors, and special cycles designed to fly around the steeply banked track. As they flash past at up to 37 mph (60 km/h), riders and machines move as one, in perfect harmony.

⬆ Made for two
This tandem was designed for cycling fast over long distances. It has a lightweight frame and its curved drop handlebars allow the rider to adopt a low, crouching position. This tandem has a twelve-speed gear unit to help with hills.

⬅ Racing cycle
Bikes such as this one are made of light metal tubing. There are no mudguards, and slim wheels reduce the load further. Toe clips stop cyclists' feet from slipping when they pedal fast.

⬋ Power racer
Racing motorbikes are the swiftest things on two wheels. They can zip around the track at over 180 mph (290 km/h)—if the rider dares! Powerful engines, streamlined bodies, and a low riding position all help them to go faster.

⬋ Pedal power

British cyclist Chris Boardman won gold at the 1992 Olympics on this space-age cycle. The bike's frame and wheels are made of a carbon composite material that is much lighter than metal. Everything is pared down —there is no extra equipment and every surface of the frame is streamlined. Special handlebars give the rider an ultra-low position as the air rushes past his streamlined helmet.

Record-breakers

A LOT OF US like traveling fast, but for a few people speed alone is not enough—they want to be the fastest in the world! Achieving this can take a lifetime of designing, building, and rebuilding, to create a vehicle that will smash that record. The same goes for world-class athletes, who dedicate their lives to perfecting their performance. Being a record-breaker takes a lot of hard work—and a lot of brilliance too!

⬇ Like a bullet

The Japanese Shinkansen, or bullet train, rushes passengers the 320 miles (515 kilometers) from Tokyo to Osaka in 2 hours 30 minutes. The latest experimental bullet trains are already traveling at well over 200 mph (322 km/h).

⬋ By jet across the desert

Thrust SSC is the first car to go faster than the speed of sound. Powered by two large Rolls-Royce jet engines, the car reached 763.035 mph (1,227.985 km/h) in the Black Rock Desert, Nevada, in 1997. At this speed, the main danger is that the car will leave the ground, turn over, and crash. The vehicle's large tail fin helps to prevent this.

⬅ Fastest under sail

The three-hulled boat *Yellow Pages Endeavour* broke the speed record for a sailing vessel in 1993. Powered by a single tall sail, its three small hulls skimmed along the water near Melbourne, Australia, at 53.57 mph (86.21 km/h).

➡ Blackbird

Designed more than 30 years ago, the Lockheed Blackbird is still the world's fastest airplane. It cruised at about 2,175 mph (3,500 km/h)—that's more than three times the speed of sound.

Glossary

aerodynamic Term describing an object that moves through the air smoothly.

clipper Fast sailing ship of a type popular in the mid-19th century, often used for carrying valuable cargoes such as tea over long distances.

fiberglass Artificial material made up of tiny strands of glass; it is valued by boat builders because it is relatively light.

Formula One A type of car racing, featuring powerful purpose-built, single-seat cars that compete in a series of Grand Prix races each year.

gravity Force that attracts objects to each other; the strength of this force depends on the mass of each object.

hull The body of a boat or ship.

jet Type of engine that provides forward thrust by producing a fast stream of exhaust gases from the other end.

km/h Kilometers per hour.

locomotive Railway engine, used for pulling trains and powered by steam, diesel fuel, or electricity.

maglev Experimental type of train that uses *magnetic levitation* to float a short distance above the track when it moves, reducing friction and increasing speed.

mph Miles per hour.

powerboat Any boat with an engine, although the term is usually used to describe high-performance boats designed to travel fast.

ram Pointed device fixed to the hull of an early warship, used for making holes in enemy ships.

sound barrier Point around the speed of sound when a marked increase in power is needed to go faster.

streamlined Term used to describe an aircraft, boat, or vehicle designed with a smooth shape so that it moves more freely through air or water.

supersonic Faster than the speed of sound.

tandem Bicycle for two riders; a tandem has two saddles and two sets of pedals and handlebars.

thrust A pushing force that drives a vehicle along.

vacuum Space containing no matter; in practice a vacuum is space from which as much air as possible has been removed.

wind tunnel A device used to pass a steady stream of air around an object such as a model of a car or airplane; used to help design more aerodynamic, streamlined vehicles.

yacht Light sailing vessel equipped either for racing or for luxury cruising.

Index

Page numbers in *italic* type refer to illustrations.